Mother Nature's Examples

Mother Nature's Examples

EXTREMELY INTERESTING NATURAL ADAPTATIONS AND THE EXTREMELY IMPORTANT HUMAN INVENTIONS THEY SPUN

• • •

Mustapha E. Njie: BS, MS, MA

ISBN: 1532986963
ISBN 13: 9781532986963

Dedication

• • •

This book is dedicated to the memory of my parents and sisters

Preface

• • •

CREDIT FOR THIS WORK GOES to everybody who helped me mold both my fluid and crystallized intelligence—my parents, teachers, wife, kids, friends, enemies, brothers, sisters, and even strangers I randomly encountered and learned from. Likewise, any blame for this work goes collectively to all of us. The ridiculous notion that all blame, in matters of this sort, should solely go to the author is also an indirect usurpation of all the credit from the work—a quintessential portrayal of arrogance and self-centeredness.

Taking all the blame, hence credit, for anything is always a false narrative because none of us would be where we are today without the multitudes that aided us along the way. So, as country singer Tim McGraw reminded us in his instructive and archival anthem, "Always stay humble and kind."

I wrote this book for two reasons: The primary reason is my belief that there are several take-home messages in this book that can make the world a better place. The secondary reason is that I am almost sixty years old and have started to seriously think about my mortality and the fleeting nature of life. So I decided to use some of my epistemological credibility to jot down something for posterity to peruse, criticize, grade, and hopefully use to advance humankind.

The idea for this book came when I was preparing a lesson plan on evolution and the adaptations that makes it possible. Any good teacher will tell you that a good lesson must start by engaging the students' attention. A good way to do this with high school teenagers is to provide them evidence of how people have monetized the knowledge you are about to introduce them to. This "what's in it for me?" attitude is pervasive among humans, especially high school teenagers. So, to engage my students, I started by telling them about some interesting natural adaptations and the money-making human inventions that came out of them. This caught the attention of the students so well that I decided to write a book about it.

I strongly believe that this book can encourage teenagers to become scientists, inventors, and entrepreneurs. It clearly shows them that the "science dealer" makes more money than the "dope dealer," without the danger of getting shot or going to jail.

Judging from the response I got from the first person to read the book, Rashana Moret, one of my eleventh-grade chemistry students, the mission for writing this book has already been accomplished. After reading the book, she said, "I learned a lot from it. Some things I read seemed unthinkable and surprising! I really have to learn to think outside of the box."

*She learned from the book, and it changed her way of thinking for the better—*this was the primary mission of this book.

The second person to read the book, David Young, an earth and space science student of mine last year, basically said the same thing Rashana said. He said, "This book gives me a new perspective, and inspires me to get educated—since education is the way—and to look around and use what is given to make what is desired."

The third person to read the book was also one of my former earth and space science students, Nakayla Mcneil. Some of what she said was, "I found your book very informative. It talked about science that people do not know about. Overall, I really enjoyed this book and would be happy to recommend it to my friends."

The fourth and final student to read the book, Edward Conway, a senior in one of my chemistry classes, said the book informed him about life and science.

Prologue

• • •

I HAVE BEEN A SCIENCE teacher in the Chicago public school system for about twenty years. I teach mostly chemistry, but also biology, environmental science, and earth and space science. I like teaching because of its potential to change peoples' lives for the better.

I asked four of my current and former students to read this book and give me an honest appraisal of its quality. Their comments are presented below.

RASHANA MORET, ELEVENTH GRADE, CHEMISTRY STUDENT

Hey Mr. Njie,

I just finished reading your book, and I really liked it. I enjoyed reading it and even shared some of the inventions with my mom. It was very interesting and eye opening. I learned a lot. I often wondered, but never thought too much of, or asked a question, or even researched, how or why certain things are the way they are, or how they even came about. I was very surprised and happy you asked me to read this book. Some things I would've never imagined came from something so simple in nature.

All of the points were very eye opening, but a few that stood out to me were 21: Dolphin sound communication and reliable tsunami warning systems. This stood out to me because it was saying how scientists now use knowledge from dolphin sounds to communicate over long distances to help warn people to get out of the way. So, of course, we don't have tsunamis in Chicago, but we do have tornados. I wanted to know how the tornado siren works. I tried to look it up, but I don't think I found the answer yet. However, I did take an educated guess until I find the answer.

Another one that stood out to me was 1: The Wrights brothers, airplanes, and birds. I would've never guessed that the invention of airplanes was inspired by the behavior of birds. This was very interesting to read. I was shocked.

Another one that also stood out was 31: The human eye and cameras with a wider field of view. I really like photography, and cameras, and taking pictures and all that. So to actually read about how scientists are improving cameras with the human eye as their model was really interesting to me.

Another very interesting one was 51: Fabrics that can change color instantaneously and the cuttlefish. This stood out to me because when I first read it, I had to reread it. It sounded like magic. Clothes that can change color? Without fading, I'm guessing. What? How? I'd definitely have to come across this again!

The last one that stood out to me was 54: The human blood-clotting system and plastic that can fix its own cracks and damages. This is so weird but interesting. This is another one that just blew my mind and sounded like magic. How would something like a cell phone, computer, or car be able to heal itself? That is so neat and weird. I would actually love it because my phone is pretty expensive, and if I drop it one time, my screen is definitely going to crack. I think I liked this one the most.

I really enjoyed reading your book. I learned a lot from it. Some things I read seemed unthinkable and surprising! I really have to learn to think outside of the box.

DAVID YOUNG, TWELFTH GRADE, FORMER EARTH AND SPACE SCIENCE STUDENT

Mr. Njie,

After carefully reading and slowly deliberating, I think that the work you have done is very interesting and informative. Though it is science related, it's full of interesting fun facts; it influenced me to give something back to life and the world. Above all, I really liked the epilogue.

It gave an insightful hypothesis on how we can further improve our one and only earth.

Finally, this gives me a new perspective, and it inspires me to get educated—since education is the way—and to look around and use what is given in order to make what is desired.

P.S. Thanks for believing in me and allowing me to read this thought-provoking book.

EDWARD CONWAY, TWELFTH GRADE, CHEMISTRY STUDENT

Hey, my name is Edward Conway. I'm a student attending Simeon Career Academy and was recommended by my chemistry teacher, Mr. Njie, to read his book, titled *Mother Nature's Examples*. The book is dedicated to the memory of his parents and sisters. I find it interesting that Mr. Nije is able to mix his memory of his parents and lifestyle growing up with an interesting view of science. One of the quotes that got my attention was, "My native tongue, Wolof, is full of aphorisms…One aphorism I grew up with states that knowledge is elusive, not because of distance, but because it is well hidden. It is right under your nose, but it's well camouflaged. This book reaffirms this aphorism." That quote opened my eyes to society as a whole. The story is very touching and reminds me of my loving parents. The book is also very informative about life and science. I enjoyed reading every sentence in the book. He used an intense form of grammar to present information that is useful to the human brain. As far as science goes, the book is extremely great, and I recommended it to my mother to read, and she loved it. Great book!

NAKAYLA MCNEILL, TENTH GRADE, FORMER EARTH SCIENCE STUDENT

Mr. Njie:

I found your book to be very informative. It talked about Science that people do not know about. For instance, in 54, when you talked about the fact that there can be a screen that repairs itself, I did not know that. I was attracted to that because I always break my phone. Another one that caught my attention was 66—the one about cats and safer thumbtacks. That is so true. I am always sticking myself with thumbtacks and to know that there is a tack that stays closed until used—inspired by my favorite animal, the cat—is fascinating. Overall, I really enjoyed this book and would be happy to recommend it to my friends.

Introduction

• • •

GROWING UP IN A DEVELOPING country in West Africa, I always marveled at how people in the developed world could come up with all the amazing inventions that made life so much easier. "How come they are so smart?" I would ask myself. In fact, one of the primary reasons that made me want to come to America was to find the answer to this question. Well, I think I found the answer—the developed countries just discovered the secret of copying nature. Almost every major human invention imitates nature. This book is going to provide ample evidence of this. The difference is that people in developed countries are better educated and are therefore better able to understand, interpret, and apply their observations. As Pasteur said, "Luck favors the prepared mind."

Adaptations are phenotypes or characteristics that help individuals adjust to changes in their environment. These individuals will then survive, reproduce, and pass on the genes that enabled them to survive to the next generation. Adaptations can be physical, like the giraffe's long neck, or behavioral, meaning something the organism does to survive, like migrate to more favorable pastures or burrow in holes until conditions are favorable.

Almost every surviving species is a success worthy of emulation because those who could not adjust to changes in their environment are long gone. The fossil record is replete with the remains of individuals that failed the test, and evolutionary biologists refer to them as evolutionary dead ends. These evolutionary dead ends pose a major problem to the proponents of intelligent design because the intelligent designer, if ever there was one, sure made a lot of mistakes along the way. Fossil evidence unequivocally shows that the history of life was one of trial and error—the very antithesis of willful intelligence and an all-knowing intelligent designer, who is above mistakes. Science, like all truths, thrives on evidence, empirical or otherwise, and all the evidence supports the Darwinian paradigm, *not* intelligent design.

My native tongue, Wolof, is full of aphorisms. My parents constantly threw aphorisms at us, reminding us of the power of education and warnings about the many pitfalls in life and how to maneuver them. One aphorism I grew up with states that knowledge is elusive, not because of distance, but because it is well hidden. It is right under your nose, but it's well camouflaged. This book reaffirms this aphorism.

Another aphorism I grew up with states that notorious individuals have white feathers. This makes them stand out in the crowd and makes them easy to pick on. In other words, they cannot blend in or camouflage, and this makes them easy targets.

Camouflage, by the way, may be the most important of all adaptations—so important that it may be the only adaptation found in all biomes everywhere! It gives real protection, as it makes one difficult to see. Both predators and prey use it to survive, and all human armies have copied it.

I tried to write this book in such a way that anybody can read and understand most, if not all of it, regardless of background. This is because another aphorism I grew up with states that "misunderstanding does not exist, only the failure to communicate."

Human inventions are like analogous structures in evolutionary biology because they are similar to natural adaptations for the same reason analogous structures are similar to each other—to overcome similar challenges. Unlike homologous structures, analogous structures are similar, not because of relatedness, but because the organisms happen to live in a similar environment and are therefore faced with similar challenges. For example, many adaptations of organisms in the hot African desert are usually very similar to those of organisms in the hot Australian desert, despite the fact that the organisms evolved millions of miles away from each other and have no substantial genetic relationship. This type of evolutionary relationship is called convergent evolution. I say "substantial" because all organisms are related at some level. This is evident from the fact that all organisms use the same DNA molecule, or deoxyribonucleic acid molecule, to store the genetic information used to make proteins and the same ATP molecule, or adenosine triphosphate molecule, to store energy.

Examples of Extremely Interesting Natural Adaptations and the Extremely Important Human Inventions They Spun

• • •

ALMOST ALL HUMAN INVENTIONS ARE either biomimetic or biomolecular products. The former copied function from nature, while the latter copied both structure and function from nature. The following are examples of extremely important human inventions that were copied from nature.

1) **The Wright brothers, airplanes, and birds**: The invention of airplanes was inspired by the behavior of birds. In fact, the Wright brothers finally succeeded in designing an airplane when they realized that birds can glide on air currents and stay in flight without flapping their wings. This, most likely, led them to the realization that anything in flight, with enough thrust or propulsion to overcome drag, will keep flying.

2) **Gecko's eyes and better multifocal cameras and lenses**: Geckos are among the very few animals that can see colors at night. The adaptation they have is a series of concentric rings in their eyes that have different refractive powers, meaning the rings bend light at different angles. This enables geckos to be able to focus light of different wavelengths at the same time, giving them a multifocal optical system. In fact, the reason large telescopes do not use refractive lenses is because human-made refractive lenses cannot focus light of different wavelengths at the same time. When they

focus short-wavelength light, long-wavelength light becomes distorted, and when they focus long-wavelength light, short-wavelength light becomes distorted. The problem is called chromatic aberration. Geckos have no problem with chromatic aberration, and lens manufacturers should take notice.

3) **Echolocation by bats, dolphins, and toothed whales and sonar and radar Technology**: Bats, dolphins, and toothed whales use echolocation, or sound waves, to locate and catch prey in dark environments with almost no visibility. Special organs in these mammals release high-energy and high-frequency sounds with relatively small wavelengths. The size of the wavelengths is important because wavelength is inversely related to both frequency and energy. This means that the smaller the wavelength, the greater the amount of energy it carries and the greater its frequency. These organisms evolved to release small wavelength, hence powerful, high-energy sounds.

Sound travels at a speed of 1,500 meters per second in water, and the speed at which the sound returns to the organism that released it, after hitting the target, tells the releasing organism the location of the target. If the target is close by, the sound returns quickly, and if the target is far away, the sound takes longer to return. The time it takes the signal to return and the distance travelled by the signal have a direct, or linear, relationship. Due to this fact, using the speed of sound in water as a constant, one can easily develop a linear algorithm to consistently and reliably predict the distance of objects based on how long it takes a sound signal to return to its point of origin. For example, if it takes one minute for a signal to return, the object is (1,500 x 60 seconds)/2 meters, or 45,000 meters, away. You divide by two to compensate for the time it took the signal to travel to the object. Only the return time is needed to determine the distance of the object. The military copied this adaptation and use it to locate objects both above and below the seas. The technologies are called SONAR, which means

Sound Navigation and Ranging, and RADAR, which means Radio Detection and Ranging.

Also, these technologies have now been incorporated into a smartcane for the blind by scientists in India. The smartcane is fitted with a device that emits ultrasound and a sensor that receives the sound wave when it returns to the cane. It works almost exactly like SONAR and RADAR on airplanes and ships, helping the blind "see."

4) **Leonardo da Vinci, birds, fish, ships, airplanes, and helicopters**: Did you know that Leonardo da Vinci, in the fifteen[th] century, may have been the first one to design ships, airplanes, and helicopters based on his observations of birds and fish in the wild? This fact is reflected in the art and writing he left behind.

5) **Leonardo da Vinci, tortoises, turtles, and armored tanks**: Da Vinci was the first to sketch what he called a mobile protected gun platform in the fifteen[th] century, based on his observations of how tortoises and turtles retracted into their hard, protective shells when confronted with danger. He said it would be safe and invulnerable to enemy attacks.

6) **Bees and a smart electrical grid**: Bee hives bustle with activity, but somehow, the bees are in sync, and when something needs to be done, the appropriate individual or individuals do it instinctively. This amazing behavior inspired Regen Energy's design of a smart electrical grid. They used wireless controllers, or sensors, to balance and transfer loads to where they are most needed at any given time. Needless to say, this tremendously maximized the efficiency of energy use.

7) **Photosynthesis and power plants with an infinite supply of energy from free, renewable resources, like sunlight, carbon dioxide, and water**: The closest humans have come to imitating the energy-producing "miracle" of photosynthesis are solar panels. However, solar panels produce less energy than it took to

make them. If we can figure out a way to make energy from readily available, free, renewable raw materials, like sunlight, water, and carbon dioxide, like plants do in photosynthesis, we would solve all of the world's energy needs in a very sustainable way. This would lead to a continuous cycle of nonstop, nonpolluting, energy production.

In 2009, scientists at the US Department of Energy's Berkley laboratory found that tiny crystals of cobalt oxide can split water in half, just like plants do during photosynthesis. In plants, the process is called photolysis because they use energy from sunlight to split water. This is a monumental achievement, but it only solves half the puzzle. The other half of the puzzle is how to use sunlight or a similar renewable source of energy to create something similar to the hydrogen or proton concentration gradient plants use to make the energy that is then stored in ATP, or Adenosine triphosphate, molecules.

When water is split during photolysis, every two molecules of water produce four hydrogen ions or protons, four highly energized electrons, and a molecule of oxygen. The four protons diffuse directly into the thylakoid membrane, and the energy in the four energized electrons is used to pump more hydrogen ions into the thylakoid membrane from the stroma. Needless to say, this creates an extremely high concentration of hydrogen ions inside the thylakoid membrane. After a while, this extremely high concentration of hydrogen ions causes the ions to readily diffuse out of the thylakoid membrane, with great force, into the stroma. In so doing, the force of the diffusing protons is used to attach phosphates in the food the organism consumed to molecules of ADP, forming ATP. This is how the sun's light energy is converted to the chemical energy stored in foods. The process is called chemiosmosis, and the enzyme involved is called ATP synthase. Figuring out this other half of the puzzle will solve all our energy needs.

This is not far-fetched because a company in Israel says it has developed glass-printed solar cells sensitive enough to generate power from indirect indoor lighting, without the need for a semiconductor, to run wireless electronics. A dye in the solar cell reacts with light, creating an electrical charge in the absence of a semiconductor. This is very similar to how chlorophyll reacts with sunlight during photosynthesis. The company is called 3-G Solar Photovoltaics, and its director, Barry Breen, said that the device will be built into wireless electronic devices, and there will be no need to ever change or recharge a battery again!

8) **Seafloor plants and wave turbines for energy generation**: Scientists in Australia designed the bioWAVE ocean wave energy system just like the little plants that grow on the seafloor. Like the plants, it sways with the movement of the ocean, and the blades capture the movement of the waves. This mechanical energy is then converted to electrical energy by the magnets and wires in a generator.

9) **Namib beetle, Fogquest nets, and the Dew Bank Bottle**: In the hot, dry Namibian desert of southwest Africa, water is hard to find and is a very precious commodity. The Namib beetle evolved an adaptation to solve this problem by collecting moisture from the fog that accumulates in the desert air when the hot sun goes down. The fog regularly accumulates because when the hot desert air of the day mixes with the cooler nighttime air, moisture precipitates as fog in the early morning. The Namib beetle collects this moisture by jerking its back upward and lowering its front end, and the fog that lands on its waxy, water-repellent wings collects as water in depressions along its wings before rolling down to its mouth. The Fogquest nets and the Dew Bank Bottles, biomimetic products now used to collect moisture in dry areas of the world, were made based on this adaptation.

These two biomimetic products, or modified versions, will soon be household items in many parts of the world because scientists predict that by 2050, water will be an extremely scarce commodity in most parts of the world.

10) **Bacteria and the cleanup of toxic waste and oil spills**: Some bacteria, like *Rhodococcus* and *Thiobacillus*, consume toxic waste and convert it to harmless products. These bacteria are now prime candidates for cleaning toxic waste dumps. They can also consume heavy metals and detoxify chlorine, aromatic hydrocarbons, and other carcinogens. Another option is to figure out how the bacteria functions during detoxification and make energy-efficient machines that perform the same function.

11) **Boxfish and cars less dependent on wheels for efficiency of movement**: One of the most perplexing observations for any inventor or scientist is the absence of wheels in nature. Almost all human locomotives use wheels to move, but no creature in nature evolved anything remotely close! Attempts to explain the anomaly are many—the simplest of which is that wheels are less complex than limbs and are easier to make. However, Mercedes Benz went further. They copied the aerodynamic design of the Boxfish and designed a car that relies less on wheels for efficiency of movement and more on its shape. The car turned out to be extremely fuel efficient because of the reduction in drag that its design confers.

12) **Our legs and legged robots**: Due to the limitations of wheels in rugged terrain, NASA is developing legged robots, inspired by the design of the human pentadactyl limb plan, for use in their missions.

13) **Cockleburs and Velcro**: In 1948, Swiss engineer George de Mestral microscopically examined the stubborn cockleburs that were attached to his clothes and his dog after a trip in the field and saw the tiny hooks they used to attach themselves. He copied the design of the hooks and invented the biomimetic and biomolecular product called Velcro, now widely used as a fastener for clothes. His invention consists of two strips of thin plastic sheets, one covered with tiny loops and the other with tiny flexible hooks. The

loops and hooks adhere to each other when pressed together and can be easily separated when pulled apart.

14) **Geckos' feet and industrial glues and adhesives**: Geckos' feet can adhere to any surface, no matter how smooth, without glue or adhesives because of setae, or tiny hairs, on its feet. The foot is made of tiny hairs that hierarchically split to smaller and smaller hairs. Each of these hairs is so small that it uses molecular, or Van der Waals, forces to stick to surfaces. To get rid of the grip and move on, they just change the direction of the hairs. Geckskin, a strong, human-made adhesive, is a biomimetic product copied from the gecko. It was developed by a group of scientists at the University of Massachusetts, Amherst. It is only sixteen square inches but can hold up a flat-screen TV.

Also, until now, climbing robots have used suction cups, which are slow and inefficient. Mark Cutkosky, a mechanical engineering professor at Stanford University, developed the "stickybot," which uses tiny fibers with split ends like the ones on the gecko's feet. His invention successfully overcame the disadvantages posed by suction cups on robots. To paraphrase Cutkosky, like the geckos, the robots hook and unhook themselves from the surface by simply changing the directions of the hooks on their feet.

15) **Kingfisher beaks and noise-pollution reduction in high-speed trains**: One of the disadvantages of high-speed trains is the amount of noise they put into the environment when they emerge from the underground tunnels because of the change in air pressure. Japanese engineer Nakatsu dealt with the problem by copying the design of the kingfisher's beak in redesigning the nose of the Shinkansen bullet trains. The significant reduction in drag that resulted not only reduced the noise levels, but also increased the trains' speeds and reduced their fuel consumption. Windmills have also adapted this design. Nakatsu was attracted to the kingfisher bird because its ingeniously designed beak enables it to dive into the water at high speeds to catch fish without making any noise.

16) **Blue mussels, glue that can work on wet surfaces, and surgery without sutures**: A fundamental problem posed by mussels to modern industry is their tendency to adhere tenaciously to any surface, using a liquid adhesive they produce. They clog up pipes and the hulls of ships, costing millions of dollars in removal annually. Scientists at MIT studied the structure of the mussel glue and made a biomolecular prototype that incorporated the proteins that make mussel glue work on wet surfaces. The artificial glue works on wet surfaces and may be used in the future by surgeons to operate without sutures as well as repair ships and other underwater installations.

17) **Robotic cars and schools of fish**: Self-driving, robotic cars must be able to detect and maneuver around obstacles. Engineers at Nissan effected this skill in their robotic car, EPORO, by studying how large schools of fish maneuver around large obstacles like corals. Like schools of fish, the robots quickly share information and change direction as needed.

18) **Self-Driving cars and pigeons**: Pigeons are known for their ability to flawlessly navigate through thick forests at extremely high speeds. Also, they always take the same way out of a forest as they came in and tend to always pick the straightest and shortest route available. Huai-Ti Lin and his team at Harvard are trying to understand the adaptation or adaptations that pigeons use to effect this flawless and efficient movement, using it to improve the "smartness" of self-driving cars. One thing we know for sure is that pigeons have three-hundred-degree peripheral vision. That is, they do not have a blind spot, like us, and can see almost all around them without turning their heads.

19) **Smart motorcycle helmets and our brains**: Markus Weller suffered a devastating motorcycle accident when he got distracted and rear-ended a car. He developed the smart motorcycle helmet, called Skully AR-1, to prevent such an accident from happening again. The helmet functions by mimicking the physiology of the human brain, and according to Weller, it will keep

riders safer by eliminating blind spots. In other words, it will give riders pigeon-like vision or better because it is like having eyes on your back. The camera in the helmet takes pictures of what is going on behind the rider and display it on a monitor in front of the rider. It is like having four, instead of two, eyes.

20) **Bats and energy-efficient government surveillance devices**: Com-Bat, a six-inch spy plane, was built by the University of Michigan for the US military. It is solar powered and has wings shaped like that of a bat. It collects huge amounts of data while running on only one watt of energy.

21) **Dolphin sound communication and reliable tsunami warning systems**: Tsunami waves do not get big until they reach the shore. To detect them early enough to warn people to get out of the way, pressure sensors that detect sound are placed under the waves in the ocean. However, sound waves travelling long distances echo and interfere with each other, reducing the reliability of the information used to warn people. Well, a warning system with a lot of false alarms quickly becomes useless because people start to ignore it. So, while looking for a solution, scientists observed that dolphins use sound to communicate over long distances without any interference problems. The dolphins overcame the problem of different sound waves interfering with each other by using sounds of several frequencies in each transmission. This eliminates interference because it is like the sound waves are travelling in different highways because of their different frequencies. A company called Evologic copied the dolphins and developed an underwater modem that is currently being used to overcome the problem of interference in the tsunami warning systems of the Pacific and Indian oceans.

22) **Whales' flippers and more fuel-efficient airplanes, ships, industrial fans, and surfboards**: Tubercles, or the bumps on the front part of a whale's flipper, improve swimming efficiency by reducing friction, hence drag. These bumps have been copied and are now prominent in airplanes, ships, surfboards, and large fans.

They increase speed and reduce fuel consumption. Biologist Dr. Frank Fish was among the first to realize the aerodynamic efficiency flippers conferred and founded a company, WhalePower, that uses this knowledge to help companies design more efficient locomotives, turbines, and fans.

23) **Eagle's wings and more fuel-efficient airplanes**: Anything that would reduce drag will increase efficiency of movement and reduce fuel consumption. One of the main causes of drag in flying airplanes are the vortices, or circular air movements, created at the tip of the wings. Engineers observed that large birds, like eagles, do not seem to have problems with vortices by their wing tips the way airplanes do. They correctly reasoned that this was because of the long feathers these large birds have on their wing tips. In flight, these feathers bend upward into a vertical position, giving the bird maximum lift with minimum wing length. Up to a point, the longer the wing, the greater the lift, but the greater the drag. The wing tips allows for more lift at a given wing length with minimal drag. Engineers copied this adaptation and added wing tips, called Winglets or Sharklets, to the wings of airplanes. This immediately led to a 10 percent reduction in fuel consumption and a noticeable improvement in the efficiency of flight.

24) **The honeybee's hexagonal honeycombs and the design of strong, lightweight structures in airplanes**: The wax used to make honeycombs requires a lot of energy from the bees. So the bees evolved an adaptation to make hexagonal honeycombs that store the most amount of honey while using the least amount of space and other resources. This hexagonal design has been copied by those who design airplanes and incorporated into the structures of airplanes. The bee's honeycomb led to the discovery that honeycomb-shaped structures have minimal density and relatively high compression and shear properties, hence their use in airplanes. Their low density means they are lightweight, and their high compression and shear properties mean that they are strong and can be

made into structures with large dimensions without the problems of cracking associated with comparable, but more brittle, materials.

25) **The honeybee's hexagonal honeycombs, the pangolin, and the design of strong, lightweight bike helmets that you don't need to hold in your hand when not in use**: Alpha helmet, made by Golem Innovation, is strong, lightweight, and retracts around your neck when not in use. Its strength and lightness were copied from the bee's hexagonal honeycombs, and its retractable features were copied from the pangolin, or scaly anteater. Pangolins are known for their retractable features. Every part of their body is retractable, particularly their long tongue. When threatened, they retract all of their features and roll into a ball.

26) **Elephants' trunks and robots that carry heavy loads**: Robots are controlled by computers, and as computer technology advances, more advanced robots are being built. German engineering firm Festo built a robot that moves heavy loads by deflating or inflating air sacs in its metal sections. This is similar to how elephants use their trunks.

27) **Migratory birds, shoal movements, and more efficient wind turbines**: Wind turbines produce clean energy, but they take up a lot of space, hence their absence in large urban areas. John Dabiri of Caltech partially solved this problem by copying the movement of shoals in the ocean. When large groups of fish and migratory birds move, their placement from each other is done to minimize energy consumption and maximize the efficiency of movement. They also move with the current, rarely against it. Dabiri's windmill placement and design was copied from the fish in the ocean, and it significantly improved efficiency both in terms of energy production and the amount of space used.

28) **Shark's denticles and microbe-free surfaces in hospitals, boats, and airplanes**: A common problem that all large organisms face in the wild is microbes growing on them. Sharks solved this problem by evolving microscopic, toothlike structures on their skin called

denticles. These denticles make it impossible for microbes to grip the shark's surface. This not only frees them from the nuisance of the microbes but also reduces drag, hence the efficiency of swimming. This is because it reduces friction with the water by making their surface smoother. NASA scientists copied the design of these denticles and called them riblets; they are now prominent features on the surfaces of boats, windmills, and airplanes. They help keep the surfaces clean, which means less drag, greater movement efficiency, and less fuel consumption. The technology has also been adapted to provide microbe-free surfaces in restaurants and hospitals. Airplanes, in particular, have a problem with drag caused by dead insects on their surfaces, and this technology will help reduce that problem.

Another possible mechanism of action for denticles was proposed by Harvard marine biologist George Lauder. To paraphrase him, the denticles create a low-pressure vortex that enable the fluid medium the object is moving in to suck it forward, enhancing thrust. This mechanism, called leading-edge vortex, has been known in insects and birds but is a recent discovery in marine organisms. A group of scientists at Harvard University and Buffalo University recently observed the phenomenon in stingrays and are trying to use the knowledge to design more energy-efficient unmanned submarines.

29) **Cold-weather polar organisms like the antarctic toothfish and the freezing of human transplant organs without freezing injury**: If you have a relative waiting for a donor organ to save his or her life, you will appreciate the importance of the adaptation I am about to discuss because one of the major problems of organ transplantation is the need to keep the organs healthy and viable as they are flown from one end of the country to the next. Freezing the organs for any length of time will freeze the blood in the organs destroying them. A potential solution to this problem is found in an adaptation polar organisms have. The toothfish and

many other polar organisms have special proteins in their blood called antifreeze proteins, or AFPs, that keeps their blood from freezing. Art Devries, a graduate student at Stanford University, was the first to discover these proteins in the 1960s. The proteins lower the freezing temperature of the blood of host organisms by binding ice crystals. This prevents the ice crystals from growing and getting large enough to freeze the blood of the organism.

Scientists have isolated these AFPs, and in 2005, a team of researchers from the United States and Israel preserved rat hearts in a solution of sterile water and fish AFPs. The hearts were cooled in a freezing solution at -1.3^0 Celsius for a day, then warmed, rinsed, and transplanted into rats as auxiliary organs. After more than two hundred transplants, the researchers concluded that their technique worked because preservation time was increased from 4 hours at 5 degrees Celsius to 21 hours at -1.3^0 Celsius. The technique also worked with pigs. Unfortunately, they did not have the money to try the technique on human transplant organs. The pharmaceutical companies did not think this type of technology would generate enough money to be worth the investment. Nevertheless, this area of cryobiology holds tremendous potential for the human organ transplant industry.

30) **Tardigrades and preserving vaccines without refrigeration**: One of the most daunting problems faced by health officials in developing countries is how to prevent the spoilage of vaccines and other perishable products in rural areas without electricity, hence refrigeration.

Well, an adaptation evolved by tiny arthropodlike aquatic organisms called tardigrades, or water bears, have come to the rescue. Tardigrades can dry out and remain dormant for a century or longer, and spring right back up when they encounter enough water. These tiny organisms evolved a mechanism called anhydrobiosis, which protects their genetic machinery—that is, their DNA, RNA, and proteins—until water resuscitates them.

Two companies, Biometrica in California and Nova Labs in England, copied this adaptation and made biomolecular products that are now used to preserve vaccines and other perishables without refrigeration or electricity. These biomolecular products are based on the mechanism of osmosis, or the fact that water molecules, left on their own, will always passively move down a concentration gradient. This means that water will always move from an area crowded with water molecules to an area less crowded with water molecules. For example, Nova Lab's "candy-coated" vaccines, most likely, preserve the vaccines by absorbing any moisture that try to accumulate in the vaccine. This happens readily and passively because the "candy" has less water than the vaccine, causing water to diffuse from the vaccine into the "candy" surrounding it. The bacteria that spoils vaccines need water to survive. Because the vaccines have virtually no water, the bacteria cannot survive, keeping the vaccines fresh and viable.

31) **The human eye and cameras with a wider field of view**: A major problem with cameras is getting everybody into the picture. Camera lenses are just not adequately curved, like the surface of the human eye. The curvature of the human eye gives it a wider field of view than any camera lens. Designing camera lenses with curved surfaces, like the human eye, was not practical because the curved surface was not able to hold all the microelectronic pieces that makes a camera work. Scientists at Northwestern University and the University of Illinois built a digital camera similar to the human eye, using a meshlike material to hold the microelectronic pieces onto the curved surface. This camera has a better focus, produces better pictures, and may one day be used to make artificial retina.

32) **The mosquito's mouth, the wood-boring wasp, and painless injection needles and neuroprobes**: Usually, you don't notice the presence of a mosquito until after the bite. The adaptation that helps their mouths stabilize and painlessly penetrate skin was

studied and copied by scientists and material engineers in Japan. The information was used to make relatively painless injection needles.

Scientists and material engineers in the United Kingdom followed the lead of their Japanese counterparts and found that the ovipositor, or egg-laying organ of the wood-boring wasp, works in the same way as the mosquito's mouth, and used the knowledge to make relatively painless neuroprobes.

33) **Spider nets and new fiber manufacturing techniques**: The strength-to-weight ratio of a spider net is greater than that of steel. Its unique chemistry enables it to stretch and absorb the impact of a flying object in a manner similar to a sponge or spring. This is why it can intercept an insect flying at full speed without rupturing. It is resistant to sunlight, rain, and wind and is also insoluble in many solvents. Scientists and inventors are exploring ways to use this adaptation to create parachute wires, suspension bridge cables, sutures, and bulletproof clothing. The problem is that spiders just do not make enough fibers for industrial use. A company in Michigan, Kraig Biocraft Laboratories, is trying to solve this problem by genetically engineering silkworms to produce spider silk. Also, in 2004, a company called Dzenis reported the development of an electrospinning technique to produce strong micro- and nanofibers inspired by the spider.

34) **Spider nets and saving birds from high rises**: Birds can see spider nets from a distance and avoid them because of the ultraviolet radiation reflected by the nets. German engineers copied this adaptation and invented an ultraviolet window coating used in high-rise windows to ward off birds, preventing bird kills.

35) **Termite mounds, internal temperature control, and the design of buildings**: Building and architectural engineers are studying termite mounds for leads on how to design more energy-efficient buildings. Termite mounds stay cool in the hot desert

because cold air enters them through the bottom, pushing the warm air at the top out. In fact, the internal temperature control mechanism of the Eastgate building, an office complex in Harrare, Zimbabwe, was designed after studying termite mounds.

36) **Snails and cooler, better-ventilated buildings in the desert**: Some Iranian students won the Biomimicry Institute's student design challenge by designing a desert house inspired by snails. To keep the building cool, they made it look like a snail, using overlapping, curvy shells. This minimized the amount of sunlight that directly hit the building—the same benefit snails get when they retract into their shells when the sun gets too hot. The inside had buffer zones to maximize natural ventilation.

37) **Snails and better bulletproof vests**: A particular species of sea snail evolved a triple-layered shell to survive predators, extreme temperatures, and pH fluctuations in the bottom of the Indian Ocean around hydrothermal vents and the tough thermophiles that live around them. A group of MIT researchers are mimicking the iron-plated shell of the scaly-foot snail to make better bulletproof vests for the military.

38) **Sea lampreys and disease detection inside the body**: Sea lampreys are small, snail-like powerful swimmers. British researchers are developing a miniature robot inspired by the lamprey to navigate our bloodstream, looking for diseases and other conditions that existing technology cannot detect.

39) **Fireflies and safer cancer treatments**: Many a time, the most effective treatment for cancer is chemotherapy. However, chemotherapy has a great disadvantage in that it destroys normal cells in the same way it destroys cancerous cells. Bioluminescence, the adaptation fireflies use to produce light that attracts mates and scares off enemies, may one day be used as markers to make abnormal cancer cells identifiable and self-destruct, without affecting normal cells. The enzyme that produces the light in fireflies is called luciferase, and the gene that produces it has been

identified, isolated, and engineered into cancer cells in a lab by a group of scientists at University College in London. The cancerous cells lit up, producing light that eventually burned and killed them.

40) **Fireflies and brighter lightbulbs**: The sharp, uneven, corrugated scales on a firefly's belly help amplify the brightness of the light they produce. This adaptation was copied and incorporated into the design of the coating of newer LED lightbulbs, and it increased their brightness by about 50 percent.

41) **The Swift bird and a swift flying robot that can fold its wings backward**: Roboswift, a microair vehicle, or MAV, is a fast-flying, information-gathering robot developed by Delft University. It was inspired by one of the fastest known birds, the Swift bird. While flying, it can fold its wings backward to reduce drag, just like the Swift bird.

42) **The jellyfish and a self-powered robot**: Due to the effortless swimming mechanism of the jellyfish, it is reasonable to assume that it is highly energy efficient. Based on this belief, the US Navy is working on a jellyfish-like robot that will power itself by splitting seawater and using the hydrogen as fuel. Such a robot will be indispensable for environmental and military missions, especially in areas too dangerous for humans to venture.

43) **Hibernating bears and fighting diabetes**: Bears eat a lot of fat and sugar, becoming morbidly obese when getting ready to hibernate for the long winter. Logic suggests that any animal that behaves like this would suffer from type 2 diabetes and other diseases associated with obesity. However, the bear does not. Biochemist Kevin Corbit of Amgen discovered that the reason for this is because bears have an adaptation that prevents them from becoming insulin resistant, the cause of type 2 diabetes in humans and other animals. In other words, unlike humans, the pancreas of the bears will continue to produce insulin as long as it is needed. On the other hand, human pancreas, after

continuously producing insulin for a long time, gets exhausted and stops producing insulin. Our cells also get fed up and stop responding to insulin. Sugar then builds up in our blood, leading to diabetes. Scientists are studying the Phosphatase and Tensin homolog, or PTEN, protein switch that bears turn off to effect this adaptation so they can find a way to adapt it to human cells. We must be careful though because mutations in this gene have led to devastating cancers.

44) **Polar bears and more efficient solar panels**: An interesting adaptation of polar bears is the extremely low amount of heat they lose through their pelts. The heat they lose is so minimal that it is impossible to see them with an infrared camera. Research has shown that polar bear hair is hollow, with a foamlike material in the middle, and that it directly absorbs ultraviolet light from the sun. Scientists studied this structure and used it to make more efficient biomimetic solar panels. Also, polar organisms may be able to minimize heat loss because the arrangement of their blood vessels keeps heat away from their extremities, where it would be lost to the environment; instead, the arrangement enables them to recycle almost all of their internal heat. Researchers have also found this adaptation in arctic turtles, like the leatherback.

In a way, human-made temperature control mechanisms in large buildings were copied from arctic organisms because the heat exchangers we use operate under the same principle as that described for the bears and leatherback turtles above—keeping heat away from the edges and recycling it.

45) **Hibernating bears and kidney dialysis**: Due to their kidney adaptations, bears can hibernate all winter, without eating or drinking, and keep toxins from building up in their bodies. The toxins that dialysis patients must remove from their blood are recycled and turned into useful metabolic substances by bears. Dr. Ralph Nelson and his group at the University of Illinois in Urbana, Champagne, found that hibernating bears avoid urea toxicity, a

big problem in dialysis patients, by using the energy of their stored fats to split urea and create amino acids, which are then used to make proteins. This recycling process is faster than the process that creates the urea, so no urea builds up in the bear's blood. Also, the calcium from the destruction of bones, another major source of toxicity in dialysis patients, is recycled and used to make new bones by hibernating bears. Scientists are trying to copy this adaptation. If successful, they will give tremendous relief to dialysis patients, who spend a good part of their lives hooked up to a machine in a dialysis unit away from home.

46) **The human ear and wireless devices that can receive Internet, radio, television, and cell-phone signals**: The human ear is a fascinating piece of engineering that converts mechanical sound waves to electrical signals that are then sent to the brain. Scientist Sarpeshkar and other MIT researchers copied the design of the human ear and made a highly energy-efficient radio chip that is faster than any other. The chip is used to produce wireless devices that can receive Internet, radio, television and cell-phone signals.

47) **Vividly or brightly colored fabrics, e-readers, and butterfly wings**: Contrary to popular belief, natural colors are not always a consequence of pigmentation alone. Sometimes, they are a product of hierarchical structural arrangements at the nanoscale level. For example, in the wings of butterflies, microscales of different sizes and densities are arranged in an overlapping fashion, just like the shingles of roofs. Most of the bright coloration of butterflies is due to light reflected by this arrangement. Fiber manufacturers are now using this knowledge to make brightly colored fabrics without dyes or pigments.

Also, Qualcomm MEMS Technologies produced the first full-color, video-compatible e-reader inspired by the hierarchical structure of butterfly wings at the nanoscale level. Because the display works by reflecting, instead of transmitting light, like pigments do, it can be read in bright sunlight with no problems. It is also more energy efficient because reflecting light is much less energy intensive than transmitting light.

48) **More efficient solar panels and butterfly wings**: Chinese scientists studying how butterflies keep their wings warm for flight in cold weather discovered tiny openings in the basal layers of the hierarchical arrangement of the butterfly's wing. These openings absorb heat and keep the wings warm for flight. So instead of the flat panels we now use, maybe solar panels with layers like the wings of a butterfly will be more efficient in absorbing and storing sunlight.

49) **Underwater robots and lobsters**: The US military is using robolobsters, or robots designed like lobsters, to detect underwater mines. The robots not only borrowed their shape from lobsters, but they also have sensors mimicking the nervous system of a lobster to detect changes in its environment.

50) **Fabrics that are sensitive to sweat and female pinecones**: The University of Bath in the United Kingdom is reportedly developing a fabric with holes that will open up when the person wearing it starts sweating in order to let air in, quickly cooling the person down by drying the sweat. This would be an imitation of the hygromorphic behavior of female pinecones. Hygromorphic means the geometry of the cones changes with the level of humidity or moisture in the air. The humidity-sensitive outer layer of the female pinecone determines whether it opens or not. Because the seeds are winged, warm, dry weather is more favorable for dispersal, and that's when the cone opens up. The scales on the cone are bilayered, and the outer half shrinks more than the inner half when dry. This makes them pull away from the cone when dry and open up. When wet, the scales in both the upper and lower halves of the cone swell shut.

51) **Fabrics that can change color instantaneously and the cuttlefish**: The cuttlefish can change color instantaneously. It

effects this adaptation using specialized cells called chromato-phores. These chromatophores have sacs that are full of differ-ent color pigments, and the cuttlefish's brain sends signals to the muscles controlling the chromatophores with regard to what color pigment to release. Researchers at the University of Bristol in England are using their knowledge of this adaptation to make clothes that can change color. This biomimetic product will be highly economical for those who do not like to wear the same thing twice as well as those who care about matching when they dress.

52) **Flexible spacesuits and caterpillars**: Spacesuits must be worn by astronauts to supply needed oxygen and protect them from the pressure and temperature extremes of outer space. However, they must be designed in such a way that the astronauts can move around and do their work. The flexible design of spacesuits was imitated from caterpillars.

53) **The lotus flower and self-cleaning exterior paint, tiles, win-dow glass, umbrella fabric, and toilets**: The lotus leaf lives in a very wet, muddy environment, yet stays remarkably dry and clean. The water simply rolls off the waxy leaf surface, taking any dirt on the surface with it. The adaptation that gave the lotus leaf this self-cleaning property is called superhydrophobicity, meaning super water hating. This adaptation makes it repel almost all the water that lands on it, and as the water rolls off, it carries dirt, microbes, and other debris with it. The adaptation was first studied and cop-ied by a German botanist Wilhelm Barthlott in 1973. He studied the structure of the wax and how it is positioned on the surface of the lotus leaf, and used the knowledge to make self-cleaning paint. Now there is self-cleaning window glass, tiles, umbrella fab-ric, and, very soon, toilets. So, anytime you hear the phrase "self-cleaning," thank the lotus plant.

54) **The human blood-clotting system and plastic that can fix its own cracks and damages**: Imagine cracks on your cell phone,

computer, car, water pipes, and so forth—all repairing themselves without you doing anything or paying a dime. Sounds like science fiction, right? Well, Professor Scott White and his team at the University of Illinois have developed plastic that can heal itself and self-repair at the macroscale level. They studied the blood clotting and other repair mechanisms of the human body after a wound, and used the knowledge to make a biomolecular plastic that does almost the same thing in case of a crack or damage.

55) **Tree frogs' feet and self-cleaning medical bandages**: Tree frogs use glue produced by pads in their feet to climb trees. Glue does not stick well to dirty surfaces, but the tree frog does not seem to have a problem with this because of an adaptation it evolved. Scientists found that the tree frog self-cleans its feet while climbing, using channels in its feet that clean away dirt and debris. This knowledge may be used in the very near future to produce self-cleaning medical bandages.

56) **Ocean bottom-dwelling sponges and multidirectional, stress-resistant space suits**: In 1969, DuPont developed a nonanisotropic, or nondirectionally dependent, fabric for the space industry that could withstand stress from all directions by copying the structure of the skeleton of a sponge called *Euplectella*, or the glass sponge. Even though its skeleton is made of a weak, brittle material like glass, it can withstand ocean currents while attached to the ocean floor. This is because of the unique hierarchical structure of the sponge skeleton at the nanoscale level.

57) **Skin grafts and the parasitic worm *Pomphorhynchus laevis***: Burns and other wounds can now be better grafted using a technique copied from the parasitic worm *Pomphorhynchus laevis*. The worm pierces its host intestines, then inflates its head to securely attach itself. The biomimetic adhesive developed from this adaptation uses needles whose tips swell up when they encounter water.

The swelling keeps the graft in place and is three times stronger than surgical tape.

58) **Our noses and electronic noses**: Electronic sensors that can detect chemicals in the air, just like your nose, are now being used for quality control and environmental monitoring in the food-processing industry.

59) **Our tongues and electronic tasters**: Our tongue is the organ we use to analyze dissolved chemicals using sensors in our taste buds. Electronic tongues, or e-tongues, with sensors that make them operate just like human tongues are now being used to monitor drugs, explosives, environmental pollution, food quality, food taste, and chemical and biological weapons.

The two great advantages of both e-tongues and e-noses are that they eliminate all the biases humans bring with them, and they can be used in conditions too dangerous for humans.

60) **Desalination plants and aquaporin**: Aquaporin is a membrane-bound protein that creates a channel in the cell membrane that allows mostly water to pass through it. This adaptation was copied and used to make a biomolecular filter used in desalination plants. Desert countries like Saudi Arabia spend a lot of energy desalinating sea water for domestic use. This invention should drastically reduce their use of nonrenewable fuels for desalination purposes and the attendant pollution.

61) **Buoyant fish and the design of submarines**: In 1680, Borelli described the design of a submarine inspired by his studies of fish and their swim bladder. He proposed mimicking the swim bladder by filling goatskin inside the sides of the submarine with air through holes that opened to the outside. Oars in the form of fishtails would propel the boat by flapping. There is no evidence that Borelli's submarine was ever built, but it did get the submarine technology ball rolling.

62) **Penguins and faster, more efficient submarines**: Flavio Noca, a professor at Western Switzerland's University of Applied Sciences,

designed a penguin-inspired underwater propulsion system. He used a ball-shaped structure and parallel arms, just like the penguin's actual shoulder and arms, to recreate the flapping motion that allows penguins to move from zero to fifteen miles per hour in less than one second.

63) **Prairies, forests, and other natural flora-fauna interactions versus our modern, unsustainable, nonrenewable resource-dependent agricultural industry**: The huge monocultures that characterize modern agriculture are unsustainable in the long run because they are totally dependent on nonrenewable, polluting chemicals. Everything that makes our modern agricultural system work is nonrenewable and unsustainable—from the fossil fuels used to pump irrigation water to the raw materials used to make the pesticides and fertilizers. A nonrenewable resource is one that you *will* eventually run out of because you are using it faster than it can be produced by nature. To compound the problem, we don't know how to make the resource—only nature can make it.

We will do ourselves a big favor by copying nature and stop growing these huge monocultures of only one type of crop, covering whole states and regions, and start mixing crops from different families in the same fields. We also need to grow more deep-rooted perennials that require less maintenance, rather than the annual crops we now concentrate on, with their high energy and labor demands. This would reduce the need for artificial chemical fertilizers because the fields are not cleared off vegetation every year, as we do now, and will also minimize the need for pesticides because pests' natural enemies, which naturally reduce pest numbers, thrive in perennial mixed cultures. This is because the flora-fauna interactions are not as simplified as in monocultures that have only one type of food. In fact, The Land Institute has shown that these mixed cultures, with their deep-rooted perennials, can produce the same yields as our modern-day monocultures while simultaneously improving the water and soil resources that future agricultural production depends on.

64) **Wide ship hulls and waterfowls**: The wide beams of the hulls of modern ships confers stability and speed and was copied from waterfowls.

65) **Cheetahs and flexible prosthetics**: Older prosthetics are heavy and uncomfortable. So when Van Phillips lost his leg in a water-skiing accident, he set out to design a better prosthetic than the one he had. He read about the adaptations of fast-moving animals, like cheetahs, and was attracted to the fact that cheetahs have extremely long tendons and ligaments for muscle attachment to their skeleton. These long tendons and ligaments acted as springs to store potential energy when extended, releasing the energy when the limbs contracted back to their normal position, pushing the cheetah forward. He used this knowledge to develop a more lightweight and flexible prosthetic, using graphite to mimic the long, elastic tendons and ligaments of the cheetah. His design revolutionized the prosthetic industry.

66) **Cats and safer thumbtacks**: One of the dangers of thumbtacks is losing one and accidentally stepping on it. Inventor Toshi Fukaya of Japan solved the problem by designing a thumbtack that remains covered until it is pushed into a wall. His design was inspired by the retractable claws of cats.

67) **Our central nervous systems and electronic sensors**: The central nervous system uses sensors called neurons to inform the brain of changes in our surroundings in much the same way that the electronic sensors in smart phones, home monitoring devices, or hands-free water faucets inform the central processing unit they are connected to of changes in their surroundings. All these amazing, "smart" devices are biomimetic products inspired by our central nervous systems.

68) **Antioxidants produced by the liver, fruits and vegetables, and cancer-fighting electronic skullcaps**: The Food and

Drug Administration (FDA) approved the use of an electronic skullcap, made by the biotech company Novocure, to treat a form of brain cancer called glioblastoma multiforme, or GBM. This cancer is resistant to other forms of treatment, like chemotherapy, but the electronic skullcap kills the cancer cells by altering their electrical configuration. This is similar to how natural antioxidants in liver cells as well as fruits and vegetables convert harmful free radicals to harmless, neutral molecules. Another big advantage of the electronic skullcap over chemotherapy is that it does not destroy healthy cells.

The only reason atoms react is to fill their outermost energy level with electrons. The atoms in free radicals have unfilled outermost energy levels. So they go around, touching and pulling on everything, to get the electrons they need to fill their outer energy levels. After filling their outer energy levels with electrons, they stop roaming around and become stable. In other words, when their outer energy levels are full of electrons, they have what is called a noble gas configuration. This means that they are now stable and act like nobles, or royalty, because they now have everything they need. However, the undisciplined and haphazard behavior of free radicals, while they were looking for electrons to fill their outer energy levels, can disrupt the orderly functioning of the cell, making them extremely harmful. As they look for the electrons they need to become stable, the random touching and pulling they do in the cell disrupts the orderly function and can lead to cancer and other lethal diseases. This is one of the reasons that eating fruits and vegetables leads to good health because these foods quickly supply the free radicals with the electrons they need to calm them down, and stop their destructive cellular rampage. A healthy liver does the same thing.

69) **Decontamination of water with cactus mucilage**: Cactus mucilage, a thick, edible liquid found in cactus leaves and stems, can

effectively cleanse water of heavy metals, dirt, bacteria, and other contaminants. The cactus uses the mucilage to store water, and because of the thickness of the mucilage, the water does not evaporate easily in the hot desert.

Cleaning dirty water by boiling it with cactus leaves has been used for centuries in countries like Mexico. The thick cactus mucilage binds the contaminants and, by the force of gravity, sinks them down to the bottom of the container, leaving clean, clear water at the top. Professor Norma Alcanter of the University of South Florida and her team are researching the mechanism of the cactus mucilage to one day make a biomimetic or biomolecular product out of it. Unlike current fossil-fuel based decontamination techniques, this would conserve energy and other nonrenewable resources and will most likely not cause pollution.

70) **Zebra Fish and autism**: Autism is a serious neurodevelopmental disorder with complex symptoms. The highly social zebra fish (*Danio rerio*) is now a popular model in neuroscience for studying human brain disorders like autism and their genetic basis. This is because the zebra fish's highly evolutionarily conserved social behavior is very similar to that of humans. Also, zebra fish embryos are transparent and develop outside the mother, making their early brain development relatively easy to observe and study.

A University of Miami study shows how the manipulation of two genes in the zebra fish plays a role in the development of autism. Their researchers are utilizing zebra fish models to understand how dysfunction in either of two genes associated with autism, SYNGAP1 and SHANK 3, contributes to the disorder. Their findings show that disrupting the expression of either SYNGAP1 or SHANK 3 genes affects early brain development in the mid- and hind-brain regions, resulting in hyperexcitable behaviors. Larvae in which the two genes were dysfunctional had different brain structures—a sure sign of delayed development, slower swimming speeds, and seizures.

Based on the above findings, it is reasonable to assume that mutations in these two genes, or genes similar to them, disrupt early embryogenesis and that these disruptions play a key role in the development of autism. This knowledge will be indispensable in designing therapies and remedies for autism, a disease that now affects about 2 percent of Americans.

71) **The central nervous system and batteries**: Batteries produce electricity through chemical reactions. When activated, the negatively charged electrons in a battery flow from the negative pole of the battery, called the anode, to the positive pole, called the cathode, thus producing electricity. In fact, the definition of electricity is electrons in motion. This mechanism of electricity generation was copied from cellular, active transport pumps, like the sodium-potassium pump of the human central nervous system.

Active transport is when the cell spends energy to move materials from an area where they are less crowded to an area where they are more crowded. The cell must spend energy for this to happen because cellular materials are like us in that they will not willingly leave the comfort of less crowdedness for the discomfort of over-crowdedness. They have to be forced.

The sodium-potassium pump is a carrier protein in the cells of the central nervous system that generate electrical signals to send to the brain by pumping potassium ions into the cell and pumping sodium ions out. For every two potassium ions pumped in, three sodium ions are pumped out. As my students would say, potassium—the one with the symbol K in the periodic table—is *kicked* in, and sodium is shown the way out. Both potassium and sodium ions are positively charged, and because only two potassium ions go in for every three sodium ions that go out, the outside of the cell is more positively charged compared to the inside. Alternatively, the inside of the cell has more negative ions, or is more negatively charged, than the outside. This electrical gradient generates the electrical signals that the central nervous system uses to communicate with the brain. This is how we feel pain, hot, cold, and everything else. There are many other pumps in living cells, and almost all of them work in the same way.

Epilogue

• • •

NANOTECHNOLOGY, THE MANIPULATION OF MATTER on an atomic, molecular, and supramolecular scale, is making most of these recent inventions possible using electron microscopes, especially scanning electron microscopes (SEM), which, unlike other microscopes, do not use light to see things. Instead, it touches things, revealing details about their nature. When materials get as small as a nanometer, that is, 10^{-9} meters (1/1000000000, or a billionth, of a meter), their behavior is quite different than at the macro level. At the nano level, materials are bumpy because of the shape of molecules, sticky because of strong electrostatic and Van der Waals forces, and shaky because they are in constant kinetic molecular motion. Being able to observe the structure and behavior of nano particles, coupled with the discovery and availability of nano materials like the fullerenes, has opened up a limitless world of innovations and products. A fullerene is a form of carbon, like graphite and diamond, but with unique properties that make it ideal for organic photovoltaics, spin-on carbon hard masks, organic photodetectors, and photoresistors. It can be used in a range of applications, from medicine to electronics, because it can behave as a superconductor or semiconductor, has extreme durability, and has extreme plasticity. That is, its properties can be easily modified. It may be the only material whose solubility, electronic character, and physical nature can be modified at will to suit the purpose of the manufacturer. The first fullerene molecule was discovered in 1985 by researchers at Rice University, and this discovery earned them the Nobel Prize in chemistry in 1996.

The only prerequisite required to take advantage of the limitless opportunities of nanotechnology is education, and developing countries should take note. There is nothing magical or mysterious about Western technological development. It is all a consequence of a good and relevant education.

Copying nature is a wise and good thing because nature always selects the most energy-efficient options and uses readily available raw materials. The problem is that our imitation of nature is incomplete because natural reactions and the adaptations they produce lead to sustainable development, unlike human industrial and technological output. Almost everything nature makes is made at ambient, or room, temperature and pressure, using water as a solvent. This is mainly because natural reactions rely on reusable enzymes that drastically lower activation energy, or energy needed to start reactions, and humans have not been able to consistently duplicate the efficiency and recyclability of natural enzymes. Another reason is that natural compounds are almost always carbon-based, and carbon has unique properties that are difficult to duplicate. For example, carbon forms unusually strong covalent bonds with itself, and this enables it to form very long chains that stay stable under ordinary conditions of temperature and pressure. This, plus the fact that each carbon atom can form up to four covalent bonds, gives carbon the ability to form an infinite variety of compounds. Humans have not been able to duplicate anything close to the properties of carbon, especially at the macro level.

Also, natural adaptations rarely, if ever, perform only one task. They almost always perform multiple functions. This conserves resources and eliminates unnecessary weight—cardinal principles of product design.

After the useful lives of natural products, everything is *recycled* by nature. Currently, the manufacturing of products invented by humans leads to unsustainable development because it causes environmental degradation and pollution. If we want sustainable development, which is the only thing that makes sense, we will increase our imitation of nature and start recycling almost everything, as all natural systems do. Making products at room temperature and pressure and using only water as a solvent may be beyond our capabilities, but recycling is something we can all do. It is how all natural systems on earth have survived since the supernova that created the earth about 4.5 billion years ago. Yes, the earth itself is a product of recycling! According to the nebular hypothesis, the earth and the rest of our solar system were created from the recycled gases, dust, and debris of a giant star that imploded and died about 4.5 billion years ago—about 9 billion years after the big bang that created the universe. Need I say more about recycling?

Recycling is the real deal, so much so that three of our most important universal laws or theories—the law of conservation of mass, the first law of thermodynamics, and Einstein's theory of relativity—are all based on this concept.

The law of conservation of mass states that matter cannot be created or destroyed. It can only be recycled from one form to another. This means that the amount of matter in the universe since the big bang is the same and cannot be increased or decreased. This is why in chemistry, every time you write an equation, you must balance it to comply with the law of conservation of mass.

The first law of thermodynamics, also known as the law of conservation of energy, states that energy cannot be created or destroyed. It can only be recycled from one form to another. As with the previous law, this means that the amount of energy in the universe is the same and cannot be increased or decreased. This is why in thermo chemistry, every time you write an equation, you must balance it to comply with the law of conservation of energy.

Einstein's theory of relativity bolstered the laws of conservation of mass and energy by finally explaining the mathematical relationship between space and time as well as the relationship between energy and mass. For the first time, scientists were able to explain how the energy produced by the big bang was converted to hydrogen atoms, hence the matter everything is made of. Yes, the elements that everything is made of are made from hydrogen atoms. So, by inference, almost everything is made from hydrogen atoms. As one of my students said, hydrogen is the big mama—the mother of everything.

Einstein deduced the speed of light, in open space, to be a constant 3×10^8 meters per second and the amount of energy in any system to be equal to the mass multiplied by the square of the speed of light in open space, or $E = mc^2$. E equals the amount of energy, m equals the mass, and c equals the speed of light. This mathematical model, usually called the energy-mass equivalence equation, gave quantitative and tangible prove that the laws of conservation of mass and energy were equivalent laws based on recycling and united by the same constant—the speed of light.

Anytime I teach Einstein's energy-mass equivalence equation, some students, usually those paying attention, will always ask the following questions: What has the speed of light got to do with anything? Why is it in the equation? Where did Einstein get it from?

I understand the reasoning behind the students' questions, and I always appreciate them because it means they are listening. To somebody without a good background in physics or chemistry, it does look like Einstein pulled the speed of light from nowhere, but he did not.

The reason for including the speed of light in the equation is that energy travels in waves, and regardless of its form, it travels at the speed of light in open space. Now, the reason the speed of light is squared is that kinetic energy, or motion energy, is directly related to mass. When you increase the speed of an object, the kinetic or motion energy produced by the object increases to a level proportional to the speed of the object squared. This is why, if you decide to double your speed while running, the distance you need to stop quadruples.

Another misconception that always comes up whenever I teach these three universal laws is that many students mistake them for laws that only apply here on earth. It takes effective and deliberate teaching to help them realize that these are universal laws. So even though the energy in the gasoline burned by your car is used or lost by the car, the heat and other substances the gasoline was turned into are still somewhere in the universe.

Those who try to explain why we do not recycle as much cite cost as a reason. However, the *cost* of not recycling is far greater. One cannot realistically put a monetary value to the benefits of recycling. How much are good health and quality of life worth? How much are clean air, water, and food worth? How much is development founded on raw materials that replenish themselves and create minimal pollution worth? If there is a panacea for our socioeconomic woes, recycling may be it. It is because of recycling that humans are the only organisms with a waste disposal problem! All other organisms recycle everything.

Humans are not recycling as much as we can because of laziness. It is much easier to just throw everything away. But guess what? Nothing goes away. It is going to come back in your water, food, or some other way, and it will be much more difficult and expensive to deal with then. We are being penny wise and dollar foolish!

Our three major environmental problems today are pollution, habitat destruction, and the loss of biodiversity. All three can be solved by recycling. The more we recycle, the more sustainable our development will be; the less we recycle, the less sustainable our development will be. It is that simple.

Another major problem that can be associated with our lack of recycling is the frequent outbreak of novel viral diseases, like Ebola and Zika. These viruses have been around since the beginning of time. What has changed is that humans are going deeper and deeper into the habitats of these viruses in the African and American jungles, looking for nonrenewable resources to make things anew. These viruses do not come to us; we go to them and bring them back to the human population. The process by which viruses, or other pathogens, leave their original hosts and transfer to humans is called *zoonosis*. Recycling will definitely reduce the incidence of zoonosis.

Let me repeat: Recycling is the closest thing we have to a panacea for all the environmental and socioeconomic ills of this world. This is because recycling helps conserve limited or nonrenewable resources while simultaneously reducing pollution to a minimum. Because it conserves resources, recycling will also minimize habitat destruction and the consequent loss of biodiversity. The main reason we destroy habitat is to get resources to make things anew. Recycling would minimize this.

Recycling reduces everyone's ecological footprint, including industry, the primary culprit in the current environmental degradation crisis.

References

• • •

1) Dawson, C., et al. "How Pine Cones Open." *Nature* 390 (1997): 668.

2) HTTP://WWW.POPSCI.COM/TECHNOLOGY/ ARTICLE/2012-07/: SPEEDOS-SUPER-FAST-SHARKSKI N-INSPIRED-SWIMSUIT-ACTUALLY-NOTHING- LIKE A SHARK'S SKIN.

3) "What Is a Fullerene?" http://nano-c.com/technology-platform/ what-is-a-fullerene/.

4) Bhushan, B. "Biomimetics: Lessons from Nature—An Overview." *Phil. Trans. R. Soc. A* 367 (2009): 1445–1486.

5) https://answers.yahoo.com.

6) Benyus, Janine M. *Biomimicry: Innovation Inspired by Nature.* 2002.

7) www.smithsonianmag.com/innovation/when animals inspire inventions 59895396.

8) Romei, Francesca. *Leonardo Da Vinci.* 2008.

9) "A Smart Helmet for Motorcyclists." 2016. Newscom.com.

10) Fish and Kocak. *Biomimetics and Marine Technology: An Introduction.* 2011.

11) "Eagle's Wings Inspire More Fuel Efficient Planes—Biomimicry." https://blog.adafruit.com/2014/90/09.

12) www.triplepundrit.com/2012/03/geckskin-device-biomimicry.

13) "Using Nature to Inspire Human Innovation." Biomimetic.pb-works.com/f/Biomimetics.

14) "Self-Healing Plastic Mimics Blood Clotting." 2014. www.bbc.com.

15) "Cuttlefish Change Color, Shape to Elude Predators." 2008. News.nationalgeographic.com.

16) Bar-Cohen, Y. *Biomimetics: Biologically Inspired Technologies.* Boca Raton, FL: CRC/Taylor & Francis, 2006.

17) https://biomimicry.org/biomimicry-examples/?gclid=CJ-MsYuY2MsCFQYIaQodspINBg#.VvNFI9IrLct.

18) "13 Designs Inspired by the Sea." Webcoist.momtastic.com/2010/12/17/oceanic.

19) Dzenis, Y. "Spinning Continuous Fibers for Nanotechnology." *Science* 304 (2004): 1917–19.

20) Eggermont M. *Biomimetics as Problem-Solving, Creativity, and Innovation Tool, CDEN/C2E2.* Winnipeg, Canada: University of Manitoba, 2007.

21) www.livescience.com/28873-cool technologies inspired by nature. html.

22) "Pigeon Vision Can Teach Autopilots Fly More Accurately." www. pitts.be/index.

23) Eadie, Leslie, and K. Tushar. *Ghosh Biomimicry in Textiles: Past, Present and Potential: An Overview.* 2011.

24) www.robaid.com/bionics/biomimicry of iron plated snail could lead to better armor.

25) Fratzl, P. "Biomimetic Materials Research: What Can We Really Learn from Nature's Structural Materials?" *J. R. Soc. Interface* 4 (2007): 637–642.

26) http://webecoist.momtastic.com/2011/01/14/brilliant-bio-design-14-animal-inspired-inventions.

27) Largestfastestsmartest.co.uk/bestanimalinspiredinventions.

28) J. F. V., Vincent. "Biomimetic Materials." *J. Mater. Res.* 23 (2008): 3140–3147.

29) "Natural Adaptations and Inventions." https://books.google.com/books.

30) Borelli, G. A. *1680: The movement of animals.* 1989.

31) Reyssat, E., and L. Mahadevan. "Hygromorphs: From Pine Cones to Biomimetic Bilayers." *J. R. Soc. Interface* 6 (2009): 951–957.

32) Vollrath, F., and D. P. Knight. "Liquid Crystalline Spinning of Spider Silk." *Nature* 410 (2001): 541–548.

33) Vendrely C., and T. Scheibel. "Biotechnological Production of Spider-Silk Proteins Enables New Applications." *Macromol. Biosci.* 7 (2007): 401–409.

34) Grip, S. *Artificial Spider Silk: Recombinant Production and Determinants for Fiber Formation.* Uppsala, Sweden: Swedish University of Agricultural Sciences, 2008.

35) Autumn K., et al. "Adhesive Force of a Single Gecko Foot-Hair." *Nature* 405 (2000): 681–685.

36) Bhushan, B. "Adhesion of Multi-Level Hierarchical Attachment Systems in Gecko Feet." *J. Adhes. Sci. Technol.* 21 (2007): 1213–1258.

37) Autumn K., et al. "Evidence for van der Waals Adhesion in Gecko Setae." *Proc. Natl Acad. Sci. USA* 99, no. 12 (2002): 252–256.

38) Sitti, M., and R. S. Fearing. "Synthetic Gecko Foot-Hair Micro/Nano-Structures for Future Wall-Climbing Robots." *Proc. ICRA'03. IEEE Int. Conf. on Robotics and Automation, Taipei, Taiwan*, 14–19 (2003). 1164–1170.

39) Davies, J., et al. "A Practical Approach to the Development of a Synthetic Gecko Tape." *Int. J. Adhes. Adhesives* 29 (2009): 380–390.

40) Koch, K., et al. "Multifunctional Surface Structures of Plants: An Inspiration for Biomimetics." *Prog. Mater. Sci.* 54 (2009): 137–178.

41) Lai, S. C. S. *Mimicking Nature: Physical Basis and Artificial Synthesis of the Lotus-Effect.* The Netherlands: Universiteit Leiden, 2003.

42) Choi, W., et al. "Fabrics with Tunable Oleophobicity." *Adv. Mater.* 21 (2009): 2190–2195.

43) Bushnell, D. M., and K. J. "Drag Reduction in Nature." *Ann. Rev. Fluid Mech.* 23 (1991): 65–79.

44) Oritsland N. A., and D. M. Lavigne. "Radiative Surface Temperatures of Exercising Polar Bears." *Comp. Biochem. Physiol.* 53 (1976): 327–330.

45) "Developing Zebra Fish Models of Autism Spectrum Disorder (ASD)."http://www.sciencedirect.com/science/article/pii/S0278584613002674.

46) "Two Knockdown Models of the Autism Genes SYNGAP1 and SHANK3 Produce Similar Behavioral Phonotypes Associated with Embryonic Disruptions of Brain Morphogenesis." http://news.miami.edu/stories/2015/05/zebrafish-research-provides-clues-on-autism.html.

47) Stegmaier, T., et al. "Bionics in Textiles: Flexible and Translucent Thermal Insulations for Solar Thermal Applications." *Phil. Trans. R. Soc. A.* 367 (2009): 1749–1758.

48) Grojean, R. E., et al. "Utilization of Solar Radiation by Polar Animals: An Optical Model for Pelts." *Appl. Opt.* 19 (1980): 339–346.

49) Kinoshita, S., et al. "Mechanisms of Structural Colour in the *Morpho* Butterfly: Cooperation of Regularity and Irregularity in an Iridescent Scale." *Proc. R. Soc. Lond. B* 269 (2002): 1417–1421.

50) Zi, J., et al. "Coloration Strategies in Peacock Feathers." *Proc. Natl Acad. Sci. USA* 100 (2003): 12576–12578.

51) Vukusic, P., and J. R. Sambles. "Photonic Structures in Biology." *Nature* 424 (2003): 852–855.

52) http://inhabitat.com/fda-approves-cancer-killing-skull-cap-to-treat-brain-tumors/20161.

53) http://www.zdnet.com/article/novocures-electric-cap-for-brain-cancer-now-fda-approved/2011.

54) http://phys.org/news/2016-03-cactus-purifies-contaminated-aquaculture.html.

55) www.theosgoodfile.com/2016/05/11/16-825-am-goodbye-to-dead-batteries.

www.ingramcontent.com/pod-product-compliance
Lightning Source LLC
Chambersburg PA
CBHW021444170526
45164CB00001B/377